F4U CORSAIR
in action

by Jim Sullivan
illustrated by Don Greer

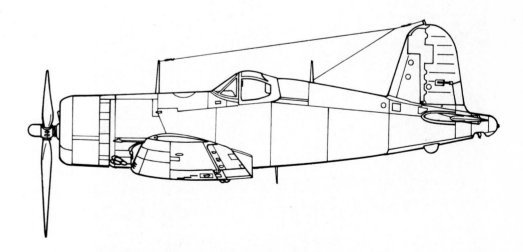

 squadron/signal publications

[Cover] A Navy F4U-5N [BuNo 124453] of VC-3 flown by Lt. Guy P. Bordelon is seen banking away from his fifth kill, an La 11, on the evening of 16 July 1953. This victory made Bordelon the only Navy ace, the only night fighter ace and the only ace to fly a propeller-driven airplane of the Korean War.

ISBN 0-89747-028-1

If you have any photographs of the aircraft, armor, soldiers or ships of any nation, particularly wartime snapshots, why not share them with us and help make Squadron/Signal's books all the more interesting and complete in the future. Any photograph sent to us will be copied and the original returned. The donor will be fully credited for any photos used. Please send them to: Squadron/Signal Publications, Inc., 1115 Crowley Dr., Carrollton, TX 75011-5010.

To: Linda
Jim Jr.
Christy

The author wishes to express his appreciation for the photo contributions and research made available by Dick Abrams, Hal Andrews, Peter Bowers, Michel Cristesco, Lou Drendel, Clay Jansson, Col. Rheinhardt Leu (USMC Ret.), Bob Lawson, Lou Markey, Paul McDaniel, Art Schoeni, Don Severson, Bill Sides, Bob Stuckey, Hank Weimer, the National Archives, Headquarters USMC, Ling-Temco-Vought and Goodyear Aerospace.

F4U-1As of VMF-113 patrol near Eniwetok, Marshall Islands. #56 is "SUNSETTER". July 9, 1944. [National Archives]

XF4U-1 Plywood mock-up. February 11, 1939. [Vought]

XF4U-1

The Chance Vought F4U was the 2nd plane to bear the name CORSAIR. The XF4U-1 was Vought's answer to the Navy design competition for a high-speed, high-altitude fighter. The Navy challenge issued on February 1, 1938 was met by the Vought design team headed by the late Rex Biesel. Just two months later on April 8th, Vought submitted a proposal. From the very beginning, the Corsair airframe was designed to mate with the new 2,000 HP radial engine from Pratt & Whitney, the R-2800, just coming off the drawing boards. After thousands of man-hours, the engineers worked out a design incorporating the inverted gull wing, the most powerful engine and the largest prop to date ever used on a fighter. On June 11, 1938 Vought received a contract from the U.S. Navy to proceed with prototypes. By February 1939 the plywood mock-up was ready for inspection by the Bureau of Aeronautics. Passing that, the XF4U-1 was assigned the Bureau number 1443 and started to become a reality. The P&W XR-2800-2 with 1805 HP for take-off was ready now and 27 months after the Navy set forth their requirements for a new fighter, Vought was ready to fly their bird. Armed with two machine guns mounted in the nose synchronized to fire through the prop and one .50 caliber in each wing, the Corsair was well provided to hold its own. In addition to this firepower, provisions were made for receptacles to house a total of 20 anti-aircraft bombs. The reasoning was for these bombs to be used by the high-flying Corsair to drop on enemy bomber formations. The all-aluminum fuselage was complimented with the inverted gull wing which along with the rudder and elevators was fabric covered. The XF4U-1 incorporated the newly designed spot-welding procedure that resulted in a low-drag smooth external finish. The wing outer panels folded upward to conserve space onboard carriers. The landing gear was designed to retract rearward with the wheels rotating 90° to lie flat within the wheelwells.

May 29, 1940 was the day the XF4U-1 first took to the air. Vought's Chief Test Pilot was at the controls as the Corsair lifted from the runway at Stratford, Conn. Even with the brief maiden flight, Chance Vought knew they had a winner. Weeks passed as Vought continued to fly and evaluate the craft. It was in October 1940 that the Corsair was turned loose on a speed course. Rear Adm. John Towers and others from the Bureau of Aeronautics watched along with Vought officials as it roared by, clocked at a record speed of 405 MPH. The Navy was understandably impressed and after favorable results from its own pilots test-flying the Corsair at Anacostia, a proposal for a production version was requested from Vought. This aircraft was to be the F4U-1. Vought engineering took over now and on December 30, 1940 the changes started that would turn the prototype into the production version. The vulnerability of the integral wing fuel tanks was recognized and they were moved to the fuselage in the production model. This change left wing space that was used to add more firepower in the form of one more .50 caliber gun in each wing. Further testing indicated that the production version would benefit greatly with additional armor-plating, even more firepower, concentrated in the wings, removal of the fuselage guns, adding a jettisonable canopy, increasing aileron area and the elimination of the wing bomb receptacles. After all the changes had been proposed, a final demonstration flight was made at Anacostia in late February 1941. The Navy issued a letter of intent for the F4U-1 and after the details were worked out, on June 30, 1941 a Navy Contract was awarded Chance-Vought to build the F4U-1. 5,559 were eventually built: 2,814 by Vought, 2,010 by Goodyear and 735 by Brewster. These figures include the F4U-1A raised cabin version.

Wing mainspars stacked by the assembly line formed the backbone of the Chance-Vought Corsair. The wing construction was virtually unchanged from the F4U-1 to the F4U-7. [Vought]

Flight testing for the XF4U-1 by Vought test pilots over the Connecticut countryside. Landing configuration has flaps and gear down, canopy in the open position. [See centerspread for color markings] May 8, 1941. [National Archives]

On October 1, 1940 the XF4U-1 set a World's Speed Record of 405 MPH while flying a record attempt between Stratford and Hartford, Conn. At this time, the Corsair was the fastest plane in the world. [Vought]

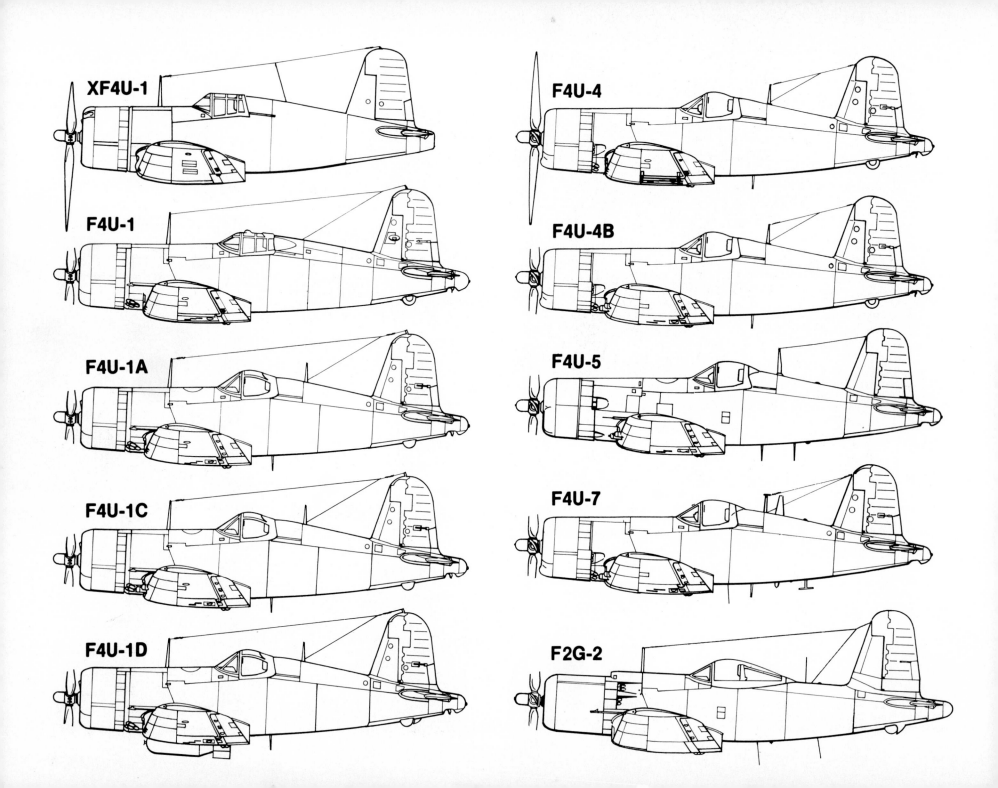

XF4U-1

F4U-1

F4U-1A

F4U-1C

F4U-1D

F4U-4

F4U-4B

F4U-5

F4U-7

F2G-2

F4U-1 Cockpit showing instrument panel and left side. [Vought]

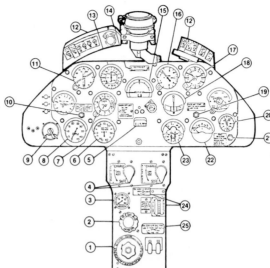

F4U-1 Forward cockpit

1. Landing gear CO2 emergency release
2. Cockpit ventilator
3. Hydraulic pressure gage
4. Gun charging controls
5. Directional gyro
6. Manifold pressure gage
7. Altimeter
8. Tachometer
9. Ignition switch
10. Fuel transfer light
11. Compass indicator
12. Armament switch boxes
13. Air speed indicator
14. Windshield defroster control
15. Gyro horizon
16. Climb indicator
17. Elapsed time clock
18. Turn and bank indicator
19. Carburetor air temp warning light
20. Fuel quantity gage
21. Fuel reserve warning light
22. Cylinder temperature gage
23. Engine gage unit
24. Switches/indicators-intercoolers and oil cooler flaps
25. Emergency oil cooler shut-off controls

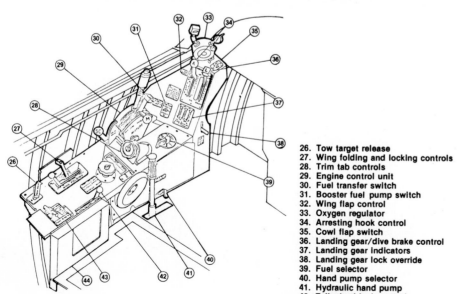

F4U-1 Cockpit left console

26. Tow target release
27. Wing folding and locking controls
28. Trim tab controls
29. Engine control unit
30. Fuel transfer switch
31. Booster fuel pump switch
32. Wing flap control
33. Oxygen regulator
34. Arresting hook control
35. Cowl flap switch
36. Landing gear/dive brake control
37. Landing gear indicators
38. Landing gear lock override
39. Fuel selector
40. Hand pump selector
41. Hydraulic hand pump
42. Tail wheel lock control
43. Manual ordnance release control
44. Map case

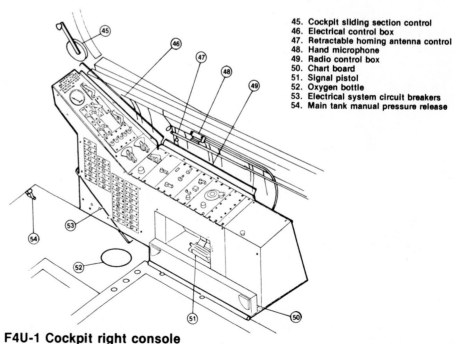

F4U-1 Cockpit right console

45. Cockpit sliding section control
46. Electrical control box
47. Retractable homing antenna control
48. Hand microphone
49. Radio control box
50. Chart board
51. Signal pistol
52. Oxygen bottle
53. Electrical system circuit breakers
54. Main tank manual pressure release

F4U-1

The Navy's initial order for 584 of the F4U-1 fighter placed Chance-Vought into full production and the Navy's confidence in the Corsair was so great that Brewster (F3A-1) and Goodyear (FG-1) were named associate contractors. The first production F4U-1 (02153) flew on June 25, 1942 and was delivered to the Navy just over a month later. The F4U-1 was rated with a maximum speed of 415 MPH, a sea-level rate of climb of 3,120 feet per minute and a service ceiling of 37,000 feet. Armament consisted of six .50 caliber machine guns and a rack on each wing just under the gun ports for a small bomb. Unacceptable performance during carrier trials forced the F4U-1 to become a land-based fighter for the beginning of its combat career. VMF-124 was the first Marine Squadron to see combat with the Corsair, but others followed soon.

[Left Above] F4U-1 Flown by Major Robert Owens of VMF-215 sports the "Spirit of 76" just forward of the fuselage insignia. Ground crewman is attaching a line to pull the Corsair from Munda's mud. August 14, 1943. [National Archives]

[Left Center] F4U-1 #12 of VF-17, the first Navy Squadron to see combat with the Corsair makes an arrested landing onboard the USS Bunker Hill during a shakedown cruise in July 1943.

[Left Below] A Marine F4U-1 of VMF-213 loses power after launch from the USS Copahee just off New Caledonia. Moments later, wetter and wiser, the pilot escaped safely. March 1943. [National Archives]

F4U-1 of VMF-222 in a close-up shot of the business end of the three .50 caliber guns. Munda Island, September 1943. [Lou Markey]

F4U-1 and F4U-1A Corsairs from MCAS Cherry Point. To compensate for the additional weight added by changes from the XF4U-1, Vought installed a P&W R-2800-8 engine with 2,000-HP for take-off. October 1943. [National Archives]

F4U-1 of VMF-215 enroute to Hawaii and the war zone. The 2nd Squadron to get the F4U, arrived at Espiritu on the 1st of July, 1943 to begin attacking the enemy bases in the northern Solomons. [Col. Leu USMC-Ret.]

Cowling
F4U-1

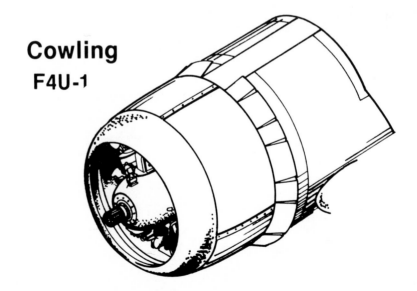

Another VMF-215 F4U-1, #8 nosed over in the mud after leaving the runway on Munda, New Georgia. Note the wide red surround on the fuselage insignia. August 14, 1943. [National Archives]

F4U-1A of VMF-212 "MARY-Jo" in a dirt revetment on Vella Lavella. From here, strikes were
flown on Rabaul harbor and in a ground support role to back up troops on Bougainville.
January 13, 1944. [National Archives]

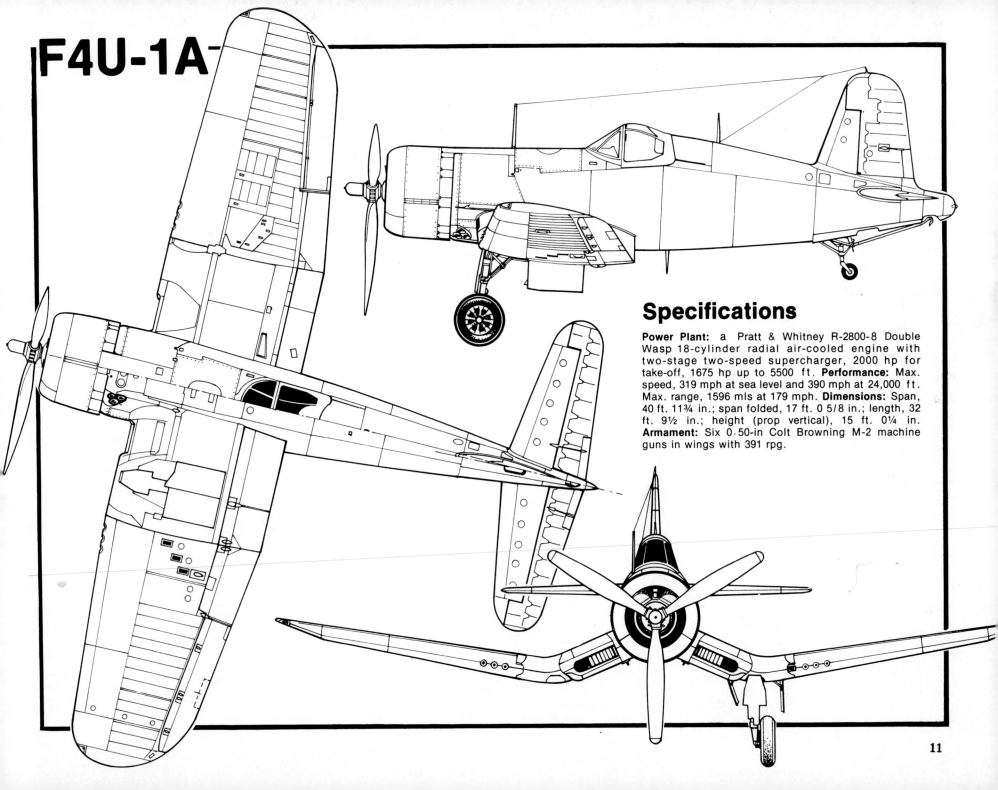

F4U-1A

Specifications

Power Plant: a Pratt & Whitney R-2800-8 Double Wasp 18-cylinder radial air-cooled engine with two-stage two-speed supercharger, 2000 hp for take-off, 1675 hp up to 5500 ft. **Performance:** Max. speed, 319 mph at sea level and 390 mph at 24,000 ft. Max. range, 1596 mls at 179 mph. **Dimensions:** Span, 40 ft. 11¾ in.; span folded, 17 ft. 0 5/8 in.; length, 32 ft. 9½ in.; height (prop vertical), 15 ft. 0¼ in. **Armament:** Six 0·50-in Colt Browning M-2 machine guns in wings with 391 rpg.

Marine FG-1A #531 displays modified raised cabin section. Problems with fuel cell leaks frequently caused taping of the fuselage skin joint areas just forward of the cockpit. September 1944. [National Archives]

Marine F4U-1A of VMF-215 on the Munda scrap pile shows evidence of cannibalization for any usable parts, a common practice throughout the war zone. September 1943. [Col. Leu USMC-Ret.]

Navy F4U-1A flown by Lt. [JG] Tom Killefer of VF-17 landed dead-stick on Nissan Island in the Green Island group after a complete engine failure. After repairs that night and following day, Killefer flew out. Note 5 kill marks under the cockpit. March 5, 1944. [National Archives]

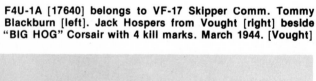

F4U-1A [17640] belongs to VF-17 Skipper Comm. Tommy Blackburn [left]. Jack Hospers from Vought [right] beside "BIG HOG" Corsair with 4 kill marks. March 1944. [Vought]

F4U-1As of VMF-113 on Eniwetok flew strikes on Wotje, Maloelap, Mille and Jaluit in the Marshalls to neutralize continued opposition there. April 11, 1944. [National Archives]

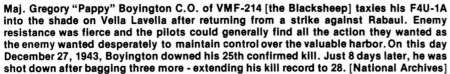

Maj. Gregory "Pappy" Boyington C.O. of VMF-214 [the Blacksheep] taxies his F4U-1A into the shade on Vella Lavella after returning from a strike against Rabaul. Enemy resistance was fierce and the pilots could generally find all the action they wanted as the enemy wanted desperately to maintain control over the valuable harbor. On this day December 27, 1943, Boyington downed his 25th confirmed kill. Just 8 days later, he was shot down after bagging three more - extending his kill record to 28. [National Archives]

Marine F4U-1A being serviced at night to enable full strength availability for morning strikes. Island sand wreaked havoc with engines. April 6, 1944. [National Archives]

Canopy Development

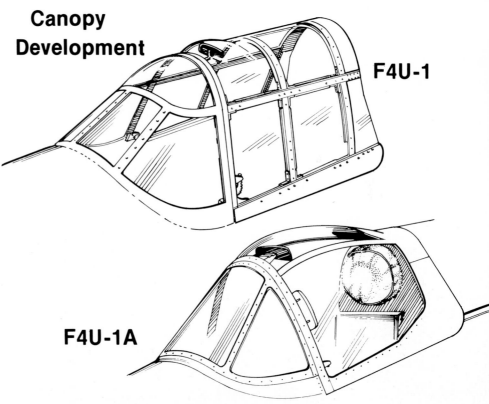

F4U-1

F4U-1A

Marine F4U-1A [50042] piloted by Lt. J.J. O'Connell of VMF-321 just prior to launch from USS Kwajalein for strikes on Rota and Pagan. Note "Hells Angels" Squadron insignia just forward of cockpit. At this point in time, the Corsair was still considered undesirable for carrier use and returned to their base in Guam after air strike. [USN Photo]

One of the most notorious of ill-fated flights occurred on January 25, 1944. After successfully launching all 24 of its Corsairs from the USS Kalinin Bay the day before, VMF-422 landed at Hawkins Field on Tarawa Atoll. On the morning of January 25, 23 of the 24 planes departed for the 700 mile trip to Funafuti with a stop-over planned for Nanomea, a distance of 463 miles from Tarawa. The Squadron lifted off at 0930 in fair weather. The flight was uneventful for a little over 2 hours until just 15 minutes out of Nanomea. Rain squalls covered them up. Dropping down to 200 feet, Maj. MacLaughlin, the C.O., passed the word for the flight to stay tight on him. Several sharp turns in the soup resulted in everyone being lost. Nevertheless, all but 3 pilots were able to stay together for a while longer. Of these 3 pilots who got separated from the others, one flew on all the way to Funafuti non-stop and landed safely there with 80 gallons of fuel left. One was never heard of again and the third managed to land on Niutao Island where he was found by friendly natives. The remaining 20 pilots and planes continued on coming into the clear over Nui Island about halfway between Nanomea and Funafuti. Two more dropped out in the water with one surviving. Now there were 18. After hitting another squall, 2 more were lost including MacLaughlin. Now there were 16. Shortly it became 15 as one more went in. Running short of fuel, it was decided that they would all ditch together while they could. Before they got into the water, two more strayed off from the others, leaving 13 that ditched together. After 2 days in the water a PBY spotted them but was damaged in the water landing and the whole lot had to be picked up by the USS Hobby. The cost in human life was enormous, 6 pilots lost their lives and of the 23 F4Us that left Hawkins Field, 22 of them were destroyed. [National Archives]

Marine F4U-1As of VMF-321 prepare for launch onboard the USS Kwajalein. "Dynamite Charley" and "Harmonica Lil's Boy" in the background. [USN Photo]

Marine Capt. Joe Moss successfully lands his F4U-1A on one main wheel. Unable to get the left gear down, Moss was able to land on Henderson Field, Guadalcanal with a minimum amount of damage to the Corsair. March 3, 1944. [National Archives]

[Right] F4U-1A [17736] of VMF-216, badly battered, made its way back to Bougainville with large hunks of control surface shot away. Despite their best efforts, Zeros were unable to down this sturdy Corsair in air combat over New Britain. January 12, 1944. [National Archives]

F4U-1A [55995] of Navy Ace Ira Kepford [VF-17] displays 16 kill marks and Jolly Roger flag on cowling. The first Corsair equipped Navy squadron to see combat, the Jolly Rogers blazed their way through the skies to bag 154 official kills during WWII. New Georgia, February 1944. [National Archives]

[Above Right] Two Marine pilots of VMF-115 demonstrate how they "piggy-backed" an Army Major in a F4U-1A. Lts. Olsen and Sharpe on two consecutive days flew bombing missions over Zamboanga, P.I., striking troops who were dug in there. The joint Army-Navy venture was successful as the Major pointed out the target, the bomb-laden Corsairs did the rest. [USMC]

Dispersed in a revetment on Bougainville, this Marine F4U-1A has just been serviced and is ready for its mission. Note the national insignia has had the borderless white bars added to the earlier circular markings. February 1944. [National Archives]

As the war raged on in the Pacific, Vought, Goodyear and Brewster were busy turning out Corsairs and by the 13th of May 1944, 3000 F4Us had been produced. As fast as they rolled off the assembly line, they were ferried out to squadrons and loaded aboard carriers for delivery to the combat zone. The war-weary earlier versions were returned by the same carriers and overhauled and assigned to a training role at bases in the States. Miramar, Santa Barbara, Jacksonville, Cherry Point, El Centro and others trained Navy and Marine pilots to fly the Corsair. The combat veteran shown above is a Marine F4U-1A being unloaded from the USS Bismark Sea at Muga Beach, California on August 7, 1944. [National Archives]

Undercarriage

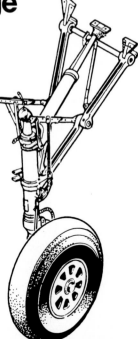

F4U-1A of VJ-5 [17580] during Field Carrier Landing Practice [FCLP]. NAS Jacksonville, Florida. February 17, 1945. [National Archives]

Marine Captain Thomas J. Ahern along with a F4U-1A assigned to a replacement training squadron at Greenville, North Carolina. Fall 1944. [Lt. Bob Hall, USMC]

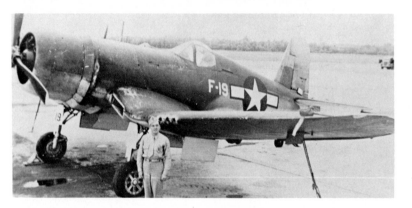

F4U-1A of Air Group 3 shows station markings "DD" assigned to Santa Barbara, California. The glossy sea blue finish was adopted in March of 1944. [USMC]

A field-engineered center-line rack made of water-pipe, metal strips bent to shape, assorted nuts and black tape holds a 500-lb. general purpose bomb. This unit was made by members of the "Flying Duces", VMF-222. Green Island, February 1944. [Lou Markey]

Brewster Bomb Rack

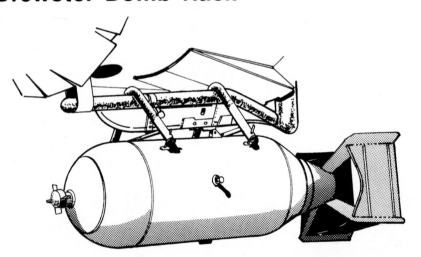

With 1000-lb. bombs hung from field-made racks, Marine F4U-1As of VMF-111 based in the Marshalls head for the enemy anti-aircraft installations on Mille atoll. March 1944. [USMC]

The fighter-bomber concept became a reality early in 1944 as members of VMF-222 worked to perfect a bomb-toting rack and release mechanism for the Corsair. Similar work was going on with VF-17. Vought factory representatives including Col. Charles A. Lindburg, Lou Markey and Ray DeLeva assisted the personnel in perfecting the rig. After thorough ground checks for strength and stability, the green light was given for flight test. Col. Lindburg, representing United Aircraft, flew the test with a 1000-lb bomb on the rack. Taking off on the Green Island bomber strip, the F4U-1A needed little extra runway to become airborne. Subjecting the rig to steep high-speed dives and maneuvers, Col. Lindburg found the Corsair handling well and proceeded to successfully release the bomb. After landing the test was given an enthusiastic thumbs-up, the modification was workable. Spreading rapidly to other squadrons, a new chapter had just opened for the Corsair. The first Marine fighter-bomber strike was flown by the "Devildogs" of VMF-111 with 8 modified Corsairs on March 18, 1944 against anti-aircraft installations on Mille atoll. Later experiments proved that Corsairs could safely and accurately deliver bombs in diving angles up to 85 degrees.

F4U-1A [18026] with a Brewster designed adapter rack securely holds 1000-lb. bomb.
Navy and Marine Squadrons used this modification to wreak additional havoc to the
already battle-tired Japanese. January 8, 1945. [Vought]

F4U-1C [50277] Experimental modification. [Vought]

F4U-1C

The F4U-1C differed from the F4U-1A in armament. In place of the 6-.50 cal. guns, Vought armed the "C" model with 4 20-mm M-2 cannons each having 20 rounds. Especially effective for ground strafing, the cannon-armed Corsair went into combat on April 7, 1945. Flying from the USS Breton, CMF-311 (Hells Belles) and VMF-441 (Black Jacks) took the F4U-1C's to Yontan Airfield on Okinawa. Vought produced an even 200 of these cannon-armed Corsairs.

Wing Development

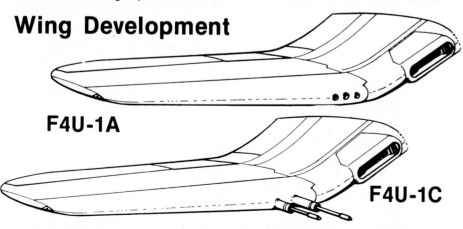

F4U-1A

F4U-1C

F4U-1C [83434] of VMF-441 off the coast of Iwo Jima. Yellow-nosed Corsair is flown by Lt. C.L. Vassey. [USMC]

F4U-1Cs of VBF-99 onboard the USS Shangri La head for combat. The white lightning bolt symbol was assigned to aircraft of CV-38. July 30, 1945. [National Archives]

Several hours before its first combat appearance, a F4U-1C of VMF-311 is spotted aboard the USS Breton. April 7, 1945. [National Archives]

F4U-1D [82332] of VMF-322, MAG-33 lands on the recently-acquired Kadena airstrip on Okinawa. April 9, 1945. [USMC]

F4U-1D

The Vought F4U-1D was the first factory-produced fighter-bomber version with a marked improvement over the F4U-1 model. Accepted by the Navy on April 22, 1944 and incorporating the increased performance P&W R-2800-8W engine with water injection, the -1D was capable of delivering 2250 hp for take-off or to escape from enemy aircraft. The water injection feature could be used for a maximum of 5 minutes in a combat-emergency situation. Twin-pylon installations were provided for carrying bombs or droppable 154-gallon fuel tanks or Napalm. In addition to these, a 3rd tank could be carried on a center-line attachment point. External rocket launching stubs were provided under each wing for 4 rockets. The upper braces in the sliding canopy were removed for better visibility in the later models. 3,862 of the -1D model were produced by Vought and Goodyear.

To repair battle damage, a F4U-1D is getting a new plate riveted on by ground crew in a service area in the Gilbert Islands. July 12, 1945. [National Archives]

Coming off the taxiway to take off strip, a Marine F4U-1D is just moments from the air. South Pacific, 1944. [Mort Hartman]

Marine F4U-1D of VMF-124/213 creates vortex on his take off roll. Finally cleared for Carrier operations, Corsairs make up for lost time in their strikes against Japanese targets. USS Essex, January 25, 1945. [National Archives]

[Above Left] F4U-1D Carries 3-drop tanks in training flight from its states-side El Centro, California base. October 14, 1944. [National Archives]

Onboard the USS Bunker Hill, a F4U-1D of VMF-451 taxies to catapult to fly against the Japanese held Ryukyus Islands. April 20, 1945. [National Archives]

[Right] A rocket-armed F4U-1D from VF-84 onboard the USS Bunker Hill takes off to join other Corsairs for a strike on Iwo Jima to support the Marine invasion. February 19, 1945. [National Archives]

Canopy Development

F4U-1D

F4U-1A

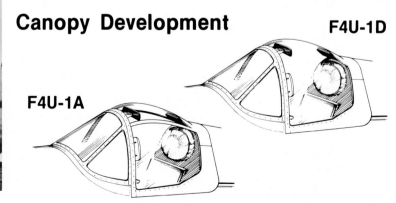

Lt. Col. Bill Millington and pilots of VMF-124/213 brief onboard the USS Essex January 1, 1945. [National Archives]

Originally designed as a carrier fighter, the F4U was not officially accepted by the U.S. Navy until late in 1944 even though the British had a 9-month head start flying from their carriers. In December 1944 VMF-124 scored another first with the Corsair when along with VMF-213 they went onboard the USS Essex. Soon, other Marine units were assigned to carriers and they included VMF-112 & 123 on the Bennington, VMF-216 & 217 on the Wasp, VMF 221 & 451 on the Bunker Hill and VMF-214 & 452 on the Franklin.

[Above Right] F4U-1Ds of VMF-124 & 213 onboard the USS Essex turn up just prior to launch, March 1945. [Vought]

[Center Right] F4U-1D from the USS Essex takes a wave-off from the LSO, January 1, 1945. [National Archives]

[Below Right] F4U-1D from the USS Essex lands onboard the USS Hancock while the Essex clears a fouled deck, June 6, 1945. [National Archives]

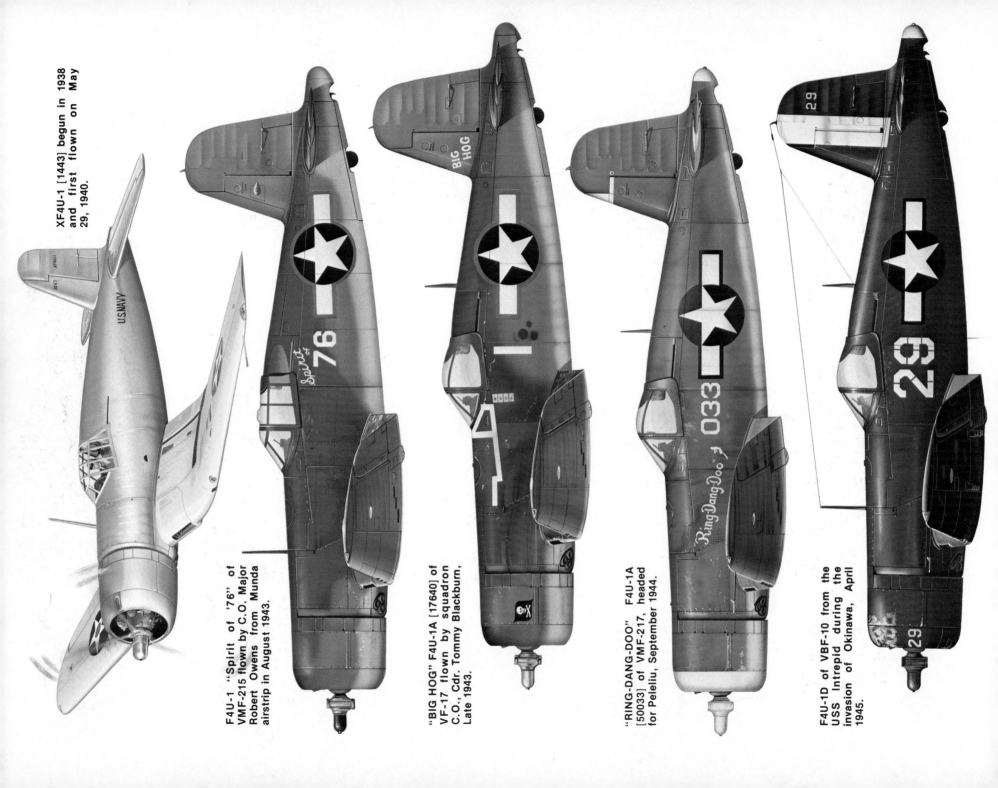

XF4U-1 [1443] begun in 1938 and first flown on May 29, 1940.

F4U-1 "Spirit of '76" of VMF-215 flown by C.O. Major Robert Owens from Munda airstrip in August 1943.

"BIG HOG" F4U-1A [17640] of VF-17 flown by squadron C.O., Cdr. Tommy Blackburn, Late 1943.

"RING-DANG-DOO" F4U-1A [50033] of VMF-217, headed for Peleliu, September 1944.

F4U-1D of VBF-10 from the USS Intrepid during the invasion of Okinawa, April 1945.

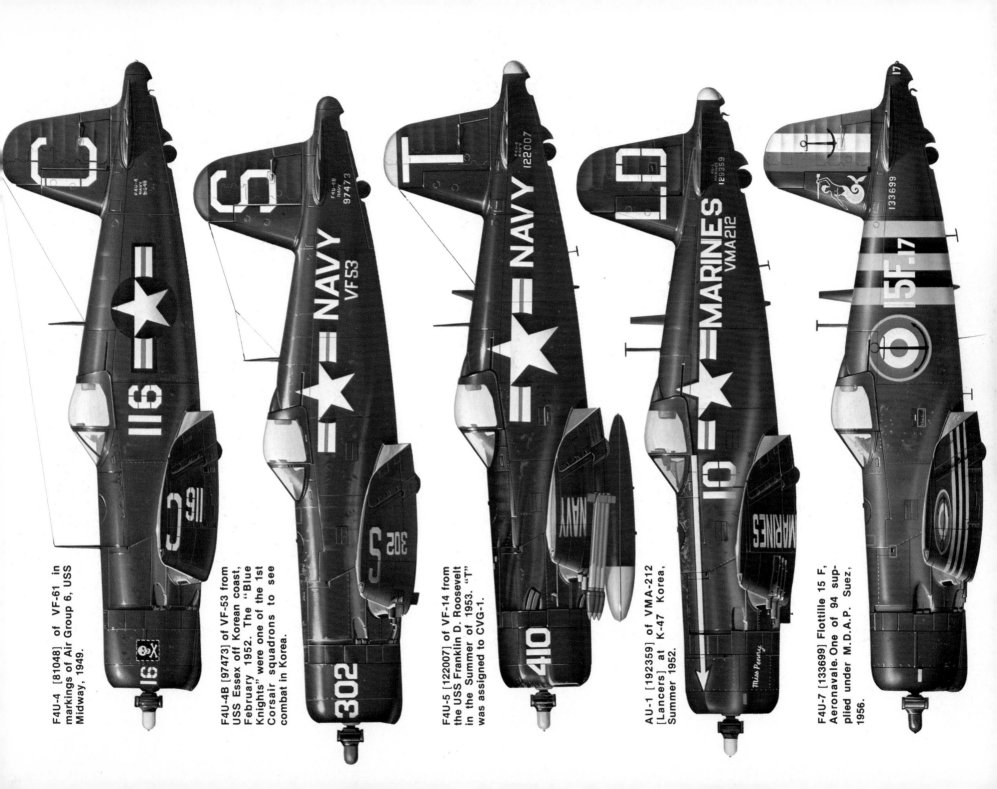

F4U-4 [81048] of VF-61 in markings of Air Group 6, USS Midway, 1949.

F4U-4B [97473] of VF-53 from USS Essex off Korean coast, February 1952. The "Blue Knights" were one of the 1st Corsair squadrons to see combat in Korea.

F4U-5 [122007] of VF-14 from the USS Franklin D. Roosevelt in the Summer of 1953. "T" was assigned to CVG-1.

AU-1 [192359] of VMA-212 [Lancers] at K-47 Korea, Summer 1952.

F4U-7 [133699] Flottille 15 F, Aeronavale. One of 94 supplied under M.D.A.P. Suez, 1956.

F4U-1D Touches down onboard the USS Ranger in December 1944. [National Archives]

F4U-1D replacement aircraft for VMF-124/213 has been off-loaded from Jeep carrier and transported by lighter to USS Essex, February 4, 1945. [National Archives]

F4U-1D of VMF-351 marked FF-61 gets the GO sign onboard the USS Commencement Bay during carrier qualifications of San Diego in February 1945. [USMC]

F4U-1Ds of VMF-221, 451 and VF-84 armed with 5-in. High Velocity Air Rockets [HVAR] onboard the USS Bunker Hill just prior to strike that marked the beginning of the invasion of Iwo Jima, February 19, 1945. [National Archives]

F4U-1D of VBF-83 is armed on hanger deck for a rocket strike on the Japanese mainland. USS Essex, March 27, 1945. [National Archives]

Fully loaded F4U-1D [57369] with 8-HVAR and 2 "Tiny Tim" 11.75-in. rockets, the largest rocket to be used by the U.S. during WWII. Tiny Tims packed 152-lbs. of TNT in the warhead and were essentially rocket-powered flying 500-lb. bombs. In mid-1944 a plan known as "Project Danny" to the Marines was formed to assign MAG-51 Corsairs the objective of attacking German V-2 rocket launching bases with the "Tiny Tims" but "Danny" never happened as the delays with the new weapon forced the U.S. and British Infantry to do the job, January 31, 1945. [National Archives]

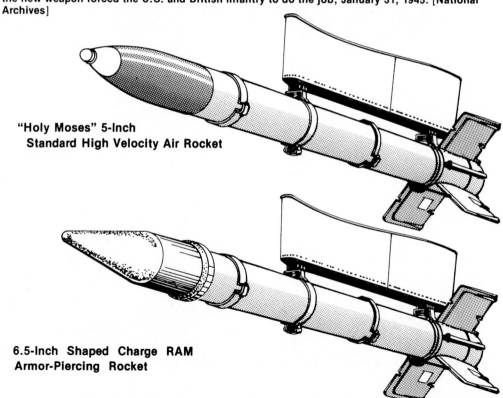

"Holy Moses" 5-Inch
Standard High Velocity Air Rocket

6.5-Inch Shaped Charge RAM
Armor-Piercing Rocket

F4U-1D of VMF-323 fires all 8-HVARs in a broadside roughly comparable to that of a Destroyer while attacking positions on Okinawa in June of 1945. [USMC]

[Above Right] Onboard the USS Hancock, a rocket-armed F4U-1D taxies forward for launch. Along with naval bombardment, rocket-launching Corsairs played a major part in softening up targets for invasion by ground troops. April 4, 1945. [National Archives]

Marine F4U-1D of VMF-312 is loaded with a pair of 500-lb. bombs and 8 "Holy Moses" rockets. Flying from Kadena airfield, these Corsairs effectively worked over the enemy positions on Ie Shima, April 17, 1945. [USMC]

Not all the Corsairs lost in combat fell to the blazing guns of enemy aircraft. Sabotage, Kamikaze attacks and ground-shelling were used effectively by the Japanese. The F4U-1D of VMF-312 sadly attests to the accuracy of the enemy gunners on Okinawa. Before the Corsair was completely cool, Marine mechanics stripped of all useable parts so that other F4Us could continue the fight. April 1945. [National Archives]

On Tacloban Field on Leyte, Marine Corsairs from MAG-12 rest in the boneyard. December 1944. [USMC]

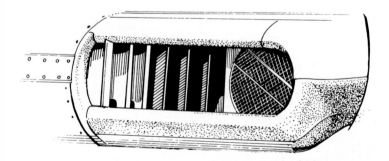

Wing Air Scoop

F4U-1D of VBF-10 flies Combat Air Patrol from the USS Intrepid during the invasion of Okinawa, April 1945. [National Archives]

Assigned to the USS Wasp, F4U-1Ds of VMF-216 and 217 struck at the Japanese mainland hitting Yokosuka and Tateyama airfields as a prelude to the invasion of Iwo Jima. The Corsair above has floated down the deck and is about to hit the barrier, February 12, 1945. [National Archives]

F4U-1D of VMF-512 catapulted from the USS Gilbert Islands, the 2nd Navy carrier to host all-Marine Squadrons onboard. VMF-512 supported the invasion of Okinawa, March 1945. [National Archives]

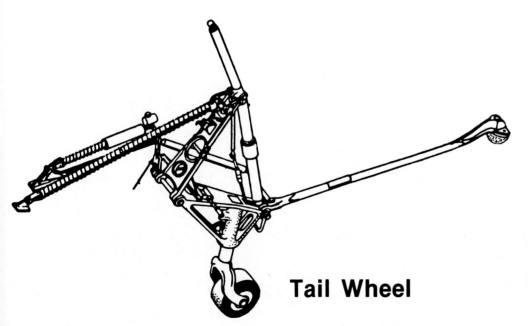

Tail Wheel

White-nosed F4U-1D comes back to the USS Bennington after a battering from Japanese defenses on Okinawa. VMF-112 and 123 were assigned to CV-20, April 7, 1945. [National Archives]

[Above Right] F4U-1D from the USS Yorktown was damaged after hitting a 5-in. gun. Note the "G" symbol in use. "RR" was assigned to CV-10 in July 1945. VBF-88. [National Archives]

[Center Right] F4U-1D in markings of USS Yorktown prior to the adoption of the "G" symbols. #41 taxies to catapult launch on July 18, 1945. VBF-88. [National Archives]

[Below Right] Getting the go-sign, another white-nosed F4U-1D leaves the USS Hancock for Japanese targets, March 21, 1945. [National Archives]

F4U-1D [82472] veered off the runway at NAB Marpi Point, Saipan completely demolishing a Navy JRB and inflicting strike damage upon itself. The pilot of the Corsair was uninjured and the JRB was unoccupied at the time, July 20, 1945. [National Archives]

F4U-1D [57632] of VF-85, the Sky Pirates, piles into the barrier onboard the USS Shangri La on December 6, 1944. [National Archives]

F4U-1D from the USS York-town in the "G" symbols assigned by the NAVY in July of 1945. [National Archives]
Other "G" Symbols:

USS Saratoga	CC
USS Enterprise	M
USS Yorktown	RR
USS Hornet	S
USS Ticonderoga	V
USS Lexington	H
USS Wasp	X
USS Bennington	TT
USS Shangri La	Z
USS Randolph	L
USS Hancock	U
USS Monterey	C

Only a few of these symbols actually were in use at the time the war ended.

After the USS Franklin was put out of action by kamikaze attacks, F4U-1Ds that were aloft landed on whatever carrier was available. Shown above is a Corsair in the markings from the USS Franklin onboard the USS Yorktown. March 25, 1945. [National Archives]

A F4U-1P of VF-84 heads out for a recon mission over Iwo Jima. February 19, 1945. [National Archives]

K-21 aerial camera is loaded into F4U-1P photo-fighter of the 4th MAW by T/Sgt. R.F. Melville. The Corsair was used in daily raids on the enemy-held Marshall Island atolls. February 1944. [Vought]

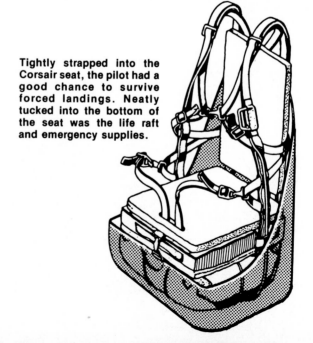

Tightly strapped into the Corsair seat, the pilot had a good chance to survive forced landings. Neatly tucked into the bottom of the seat was the life raft and emergency supplies.

F4U-1P

A field modification rather than a factory model, the photo reconnaissance version of the Corsair was essentially a F4U-1 that allowed the use of a K-21 aerial camera fitted to a special mount in the lower rear section of the fuselage. Controlled from the cockpit, this unit allowed the pilot to participate in a strike and record the results on film. Shortly after the Corsair returned to base the results of the effort were photographically available. Although the Marines are given credit for the development, both the USMC and NAVY used the F4U-1P modification.

F4U-2

In November of 1941, the Navy requested Chance-Vought to provide a radar-equipped night fighter version of the F4U-1. Vought responded with a proposal for the F4U-2 and set to work on the project. Although a mock-up was constructed, the demand of the war effort to produce F4U-1 forced Vought to delay the F4U-2 project. The Navy then, in full cooperation with CV, assigned the project to the Naval Aircraft Factory and it was there that 32 of the 34 F4U-2 Corsairs were produced, the main changes to the F4U-1 being the addition of a starboard wing radome (weight compensation being the removal of the outboard .50-cal. gun on the starboard wing), radar/transmit/receive equipment in the radio compartment, a radio altimeter system and an increased output generator. To eliminate the exhaust flames for night fighting, flame dampeners were added to the six exhaust stacks on the P&W R-2800-8 engine. The remaining two F4U-2s were field-modified by VMF(N)-532 in the South Pacific. During the operational life of the F4U-2, three units were equipped with them, VMF(N)-532, VF(N)-75 and VF(N)-101. This pioneer night-fighter developed the technology and paved the way for night-fighter tactics during WWII.

Wing Development

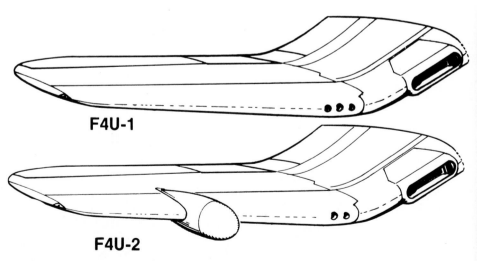

F4U-1

F4U-2

Night-Fighting F4U-2s of VF[N]-101 are spotted for launch onboard the USS Enterprise. Externally, the -2 was easily identified by the radome on the starboard wing. January 1944. [National Archives]

F4U-2 from VF[N]-101 comes up the #2 elevator of the USS Enterprise. Along with VF[N]-75, these Night-Fighters also operated from the USS Essex and Intrepid. January 20, 1944. [National Archives]

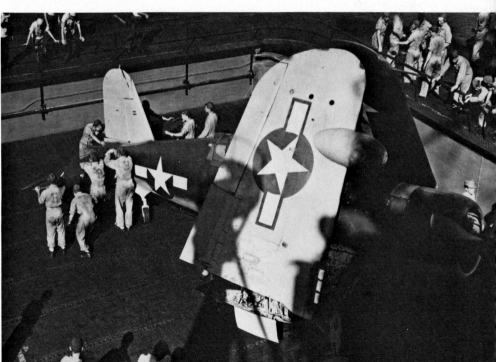

F2G-1D [88454] NATC Patuxent River, MD May 1947. [Warren Bodie]

F2G

Late in the war, Goodyear was approached by the Navy to build a high-performance version of the Corsair. Their reply was the F2G. This bubble-canopy version was to be built in two variants, the F2G-1, a land-based version with manually folding wings, and the F2G-2, a carrier version with hydraulically folding wings and arresting hook. Powered by the P&W R-4360-4 Wasp Major engine rated at 3000 HP, the F2G possessed a rate of climb that permitted it to reach 30,000 ft. in only 4 minutes. Armament was identical to that of the F4U-1D/FG-1D. Including the "X" models, 18 F2Gs were completed and flown by Goodyear before the contracts were cut back and the war was over. The F2G-1D featured an auxiliary rudder that is clearly visible in the photo above.

[Above Right] FG-1A [14092] was modified by Goodyear to test the performance of the bubble canopy stream lined, cut-down rear fuselage. Late 1944 [Goodyear].

[Center Right] XF2G-1 [13471] with the P&W R-4360-4 Wasp Major engine was the first experimental model. Late 1944. [Goodyear]

F2G-2 was the carrier-version meant to be the answer to the kamikaze menace but the war came to a close before it got into full production and only 5 of the F2G-2's were built, 1945. [Goodyear]

F4U-4 [81898] in the USMC aircraft pool on Okinawa In June 1945. [Peter Bowers]

F4U-4

The Navy's final acceptance of the first production F4U-4 (80759) came on October 31, 1944 and production of the first of the four-bladed prop Corsairs began. The F4U-4 was powered by P&W R-2800-18W engine (R-2800-42W in later -4 models) and rated as a 451 mph fighter. In addition to the 4-bladed prop and chin scoop, the F4U-4 had a redesigned cockpit to provide greater efficiency, visibility and comfort. 2,357 units were produced by Vought and although arriving late in WWII some were used by both the USN and USMC for the 4 months from May 1945 'til the end of the war. 2050 including the 5 "X" models were produced as the F4U-4 and armed with 6-.50 cal. machine guns, 297 F4U-4B models armed with 4-20mm cannons, 1 (97361) F4U-4N night-fighter with radar and 9 F4U-4P models for photo-reconnaissance. The final F4U-4 rolled off the Vought assembly line in August 1947.

F4U-4 flying in the markings of the USS Lake Champlain on June 23, 1945. [National Archives]

F4U-4 in post-war markings with the new flat bulletproof center-panel windscreen and blown rear canopy for better visibility, March 1, 1945. [National Archives]

Marine F4U-4s make show-of-force fly-over at wars' end. The main USMC fighter, the Corsair participated in campaigns from Guadalcanal to Tokyo. [USMC]

Cowling Development

F4U-1

F4U-4

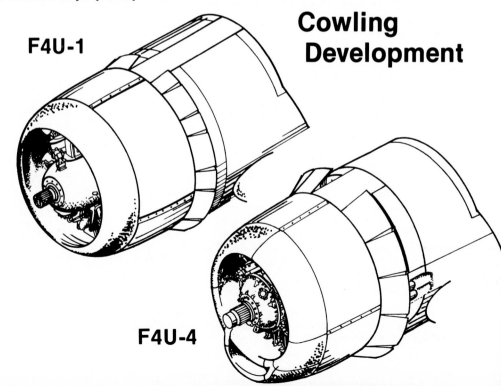

F4U-4 [82125] Stateside still carrying the Skull-and-Crossbones markings of VF-17.
Note the gun ports taped over to reduce drag, also the new blown canopy. VF-17 last
saw combat in the F4U-1A model. [Robert O'Dell]

F4U-4 [81388] flown by Lt. T.L. Clary of VF-671 lands onboard the USS Tarawa in the Mediterranean. January 19, 1952. [National Archives]

Logging over 64,000 combat sorties during WWII, the Corsair in the hands of Navy and Marine pilots was credited with the destruction of 2,140 enemy aircraft in air combat. The kill ratio of 11:1 was achieved with the loss of only 189 Corsairs although another 1,435 were lost as a result of operational accidents and enemy anti-aircraft fire.

Canopy Development

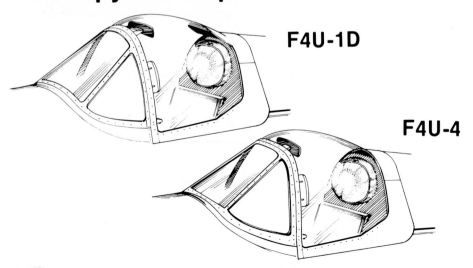

F4U-1D

F4U-4

F4U-4B cannon-armed Corsair of VF-42 [Green Pawns] heads for deep six off the deck of the USS Randolph on Feb. 22, 1947. Pilot was uninjured. [Lou Markey]

F4U-4 assigned to VMR-352 at MCAS Cherry Point, N.C. as a squadron hack. This shot in Wilmington, N.C. in 1948. [Paul McDaniel]

F4U-4 [97302] of VF-81E based at NAS Miami, Florida. circa 1949 [Robert O'Dell]

The "Weekend Warriors", the Navy and Marine reserve units flew the F4U-1D, F4U-4, F4U-5 and AU-1 versions at one time or another from the end of WWII 'til 1957. During the Korean conflict, 8 of these units were called to active duty and served honorably onboard carriers along with regular units. During peacetime, the common identifying marking on Reserve Corsairs was the bold international orange band on the rear of the fuselage.

FG-1D [92088] of VF-882 from New Orleans seen at NRAB Minneapolis, Minn. circa 1950 [Bob Stuckey]

AU-1 [129343] attached to the Naval Air Reserve Training Unit at Minneapolis, Minnesota, January 1956. [Bob Stuckey]

NAVAL AIR RESERVE STATION CODES

ANACOSTIA	A	GROSSE ILE	I	NORFOLK	S
AKRON	L	JACKSONVILLE	F	OAKLAND	F
ATLANTA	B	LOS ALAMITOS	L	OLATHE	K
BIRMINGHAM	T	MEMPHIS	M	SEATTLE	T
COLUMBUS	C	MIAMI	H	SOUTH WEYMOUTH	Z
DALLAS	D	MINNEAPOLIS	E	SPOKANE	N
DENVER	P	NEW ORLEANS	X	ST. LOUIS	U
GLENVIEW	V	NEW YORK	R	WILLOW GROVE	W
		NIAGARA FALLS	H		

F4U-4B of VMF-214 [Blacksheep] ready for pre-dawn launch for Korean strike. Armament consists of 8-HVARs, 1 500-lb. bomb and a Mk 5 150-gallon drop tank. USS Sicily, December 16, 1950. [USN-Bob Lawson]

Korea

The Korean hostilities began on June 25, 1950 and shortly thereafter the Corsair was called back to action. VF-53 and 54 onboard the USS Valley Forge hit Pyongyang, North Korea in early July and the air war was on. The Valley Forge was joined shortly by the USS Philippine Sea with VF-113 and -114 and the USS Sicily and USS Badoeng Strait with VMF-214 and -323 with others on the way. The F4U was back in action. One Marine pilot, Capt. Jesse Folmar from VMA-312 downed a MIG-15 and a VC-3 pilot, Lt. Guy P. Bordelon became a night-fighter ace. Throughout the 37 months of Korea, Corsairs participated in the action right up until the final day, which came on July 27, 1953.

F4U-4 of VF-33 going to the hanger deck onboard the USS Leyte operating off Korea, November 11, 1950. [National Archives]

42

NAS Jacksonville, Florida going full steam to refurbish F4U-4s brought out of storage for action in Korea, 1950. [National Archives]

F4U-4B [62924] of VF-113 returns to the USS Philippine Sea after mission over Korea, December 7, 1950. [National Archives]

F4U-4 [81673] of VF-821 [New Orleans Reserve] onboard the USS Princeton and Task Force 77 off the coast of Korea on July 6, 1951. [National Archives]

F4U-4 of VF-791 flying from the USS Boxer heads for the deep after a faulty catapult launch. Pilot escaped, May 19, 1951. [National Archives]

F4U-4 of VF-24 onboard the USS Valley Forge gets the turn-up signal prior to launch. Moments later the bomb-laden Corsair was airborne toward Korea, December 1950. [National Archives]

Wing Development

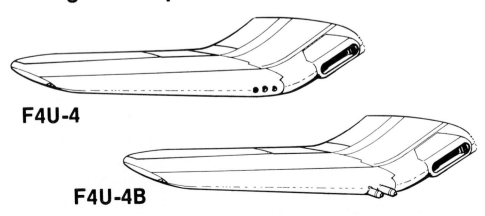

F4U-4

F4U-4B

F4U-4 of VMF-312 armed with 8, 100-lb. bombs and 1,500-lb. bomb gets the go-sign on the USS Bataan off Korea, May 19, 1952. [National Archives]

[Above Right] F4U-4B [62979] armed with 8-HVARs and a napalm tank belongs to VMF-323. Seen taking off for close air support mission in Korea, 1952. [USMC]

[Center Right] F4U-4B of VMA-332 lands onboard USS Bairoko after the squadron's last Korean sortie. July 27, 1953. [USMC]

[Below Right] F4U-4 [81865] of H&MS-33 at K-8 Kunson, Korea, 1952. [Don Severson]

F4U-4B of VF-53 "Blue Knights" leaves the USS Essex with bombs and napalm headed for Korea, March 3, 1952. [National Archives]

F4U-5 [121928] of VF-44 at Wilmington, N.C., 1949. [Paul McDaniel]

F4U-5

The first production F4U-5 flew from the new Vought plant in Dallas in April of 1946. The main external difference from previous versions was the air intake scoops at the 4 & 8 o'clock positions in the engine cowling. A second look at the F4U-5 would reveal that for the first time since production began, the outer wing panels were metal covered. Powered by a P&W R-2800-32W double supercharged engine, the -5 version developed 2,450 HP and could do even better with the use of the emergency water-injection system. Armament consisted of 4 20mm M-3 cannons and provisions for 8-HVARs and 2 center-section bomb racks. In the total of 568 units produced, 223 were F4U-5, 214 were F4U-5N, 101 were F4U-5NL (the winterized version) and 30 were F4U-5P, the long-range photo-reconnaissance version.

F4U-5 [122005] of VX-3 at NAS Atlantic City, N.J. in July of 1951. [USN-Hal Andrews]

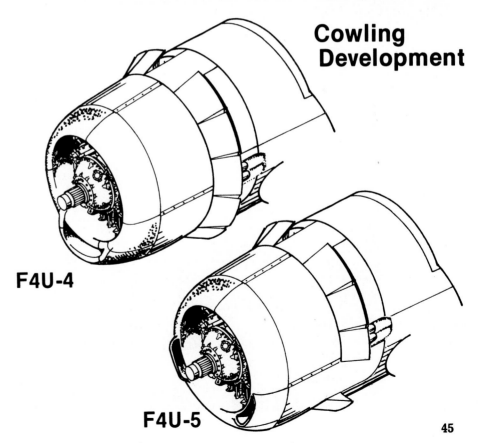

Cowling Development

F4U-4

F4U-5

45

F4U-5P [122168] of VMC-1 flying from the USS Leyte. Note open camera port under the bar in the National insignia, December 1949. [USMC]

Too late for WWII, the F4U-5 series first entered combat with the Marines during Korea. The bitter winters there forced the development of the F4U-5NL by Vought, B.F. Goodrich Rubber Co. and Hamilton-Standard. De-icing boots had never been fitted to a fighter aircraft before, but the F4U-5NL was a first. Rubber boots were installed on the outer wing panel leading edge, horizontal stabilizer and fin and propeller blades. The reality of a F4U-5NL meant that the USN and USMC had a fighter that could fly and fight even in the frigid Korean winters. F4U-5NL (124665) was converted by Vought to the XAU-1 late in 1951.

F4U-5 [121806] of VF-13 flying from the USS Franklin D. Roosevelt in Summer 1953. This Corsair was marked with medium blue on the prop hub, rudder tip, tail cone and wing tips. [Bill Sides Collection]

F4U-5N [124484] of VC-4 has just landed onboard the USS Antietam in the Caribbean Sea. The F4U-5N and -5NL series had exhaust flame-dampners to keep from impairing the pilot's night-vision, March 5, 1954. [National Archives]

Wing Development

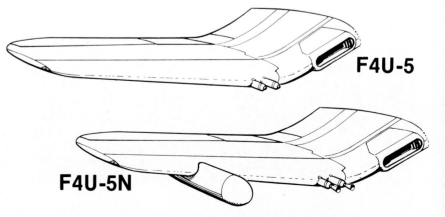

F4U-5

F4U-5N

AU-1

Built especially for the low-level ground-support role, the AU-1 was utilized in Korea by the Marines. Heavily armor-plated in vulnerable areas, the AU-1 was externally similar to the F4U-5 but without the twin cowling air intakes. These were faired over when the oil cooler was relocated to be less vulnerable to ground fire. Powered by the P&W R-2800-83W, the AU-1 had a single-stage single-speed supercharged engine. Armament was identical to that of the F4U-5 except that two additional outer wing racks were added bringing the total now to 10. This permitted the AU-1 to deliver 10 HVARs or 10 100-lb. bombs in addition to the center-section racks. By the Summer of 1952 Vought completed the run of 111 Corsairs specifically designed for low level attack. After Korea many of the combat-veteran Corsairs were assigned to Naval and Marine Reserve units.

AU-1 [129417] of VMA-212 at K-3, Pohang, Korea in 1953. Note the 5 underwing racks for HVARs or 100-lb. bombs, 1953. [Clay Jansson]

AU-1 of VMA-323 packs 6 250-lb., 1 500-lb. and 1 1,000-lb. bomb for close air support, February 20, 1953. [USMC]

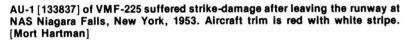

AU-1 [133837] of VMF-225 suffered strike-damage after leaving the runway at NAS Niagara Falls, New York, 1953. Aircraft trim is red with white stripe. [Mort Hartman]

Corsairs In Service With Other Countries

F4U-7 [133727] Assigned to Flottille 12F which participated in the Indochina hostilities, 1954. [Michel Cristesco]

F4U-7

The first F4U-7 was flown by Vought in July 1952 and was very similar to the earlier F4U-4B in appearance. Powered by the P&W R-2800-18W engine, the F4U-7 was produced exclusively for France under the M.D.A.P., a total of 94 Corsairs were delivered. Along with a few AU-1s from the USMC the French equipped 4 units and fought with the F4U-7s in 3 conflicts; Indochina, Suez and Algeria. The last F4U-7 rolled off the Vought Dallas assembly line in January of 1953. Late in 1964 Flottille 14F phased out the last of the French Corsairs.

F4U-7 [133663] of Flottille 14F at Cuers Naval Air Base in France, August 22, 1963. [Michel Cristesco]

The English Royal Navy during WWII equipped their Fleet Air Arm with the F4U-1, -1A and -1D Corsairs. Of the 2,012 Corsairs supplied, the English armed 19 squadrons. [Above] First to operate Corsairs off carriers, a Royal Navy F4U-1D [KD164] 1850 Sqn. taxies onboard HMS Vengeance in June 1945. [C.H. Wood]

Canopy Development

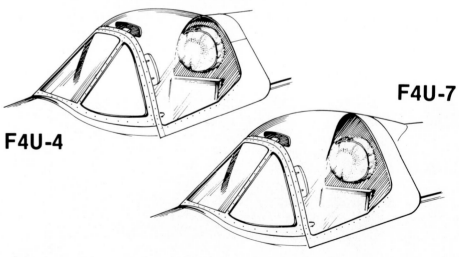

F4U-4

F4U-7

The late 1950s and early 1960s found Corsairs operating with the Argentine Navy in the form of two attack squadrons. The U.S. supplied F4U-5 and F4U-5N Corsairs which flew from the Argentine carrier Indepencia. Shown above is a F4U-5N #208 of the 2nd Attack squadron, 1960. [Roger Besecker]

During the late 1950's, the United States agreed to sell a small quantity of FG-1D and F4U-4 Corsairs from the storage facilities to San Salvador in Central America. Although phased out of service now, the Corsairs were active 'til late in the 60's. FG-1D in the markings of the San Salvador Air Force awaits delivery at El Paso, Texas in July of 1958. [Peter Bowers]

The Royal New Zealand Air Force received 419 Corsairs [231 F4U-1, 129 F4U-1A, 59 F4U-1D] and equipped 13 squadrons with them from March of 1944 'til the end of WWII, late 1944. [Vought]

Honduras was the third Latin American country to fly the Corsair. F4U-4 and F4U-5 versions were flown by them and were in service like those of San Salvador 'til the late 60's's. Rebuilt in the U.S. by surplus dealers these F4Us were among the last to see service in military control. Above is a F4U-4 in Honduran markings awaiting final acceptance at Hawthorne, California in 1960. [Ed Maloney via Tom Piedimonte]